# Understanding GENERATIVE AI



Christopher Harris

PowerKiDS press

Published in 2025 by The Rosen Publishing Group, Inc.
2544 Clinton Street, Buffalo, NY 14224



First Edition

Editor: Greg Roza
Book Design: Michael J. Flynn

Photo Credits: Cover, p. 17 Gorodenkoff/Shutterstock.com; cover, pp. 1, 3–24 (coding background) Lukas Rs/Shutterstock.com; p. 4 Andrey Suslov/Shutterstock.com; p. 5 SuPatMaN/Shutterstock.com; p. 7 https://commons.wikimedia.org/wiki/File:ELIZA_conversation.png; p. 9 FellowNeko/Shutterstock.com; p. 11 MollieGPhoto/Shutterstock.com; p. 12 Shutterstock.AI/Shutterstock.com; p. 13 Stockbym/Shutterstock.com; p. 14 pikselstock/Shutterstock.com; p. 15 gorillaimages/Shutterstock.com; p. 19 Thaspol Sangsee/Shutterstock.com; p. 21 Shutterstock AI Generator/Shutterstock.com; p. 22 Sansoen Saengsakaorat/Shutterstock.com.

Library of Congress Cataloging-in-Publication Data

Names: Harris, Chris G. (Chris Goings), author.
Title: Understanding generative AI / Chris Harris.
Other titles: Understanding generative Artificial Intelligence
Description: Buffalo : PowerKids Press, [2025] | Series: Spotlight on kids
    can code | Includes index.
Identifiers: LCCN 2024026548 (print) | LCCN 2024026549 (ebook) | ISBN
    9781499450033 (library binding) | ISBN 9781499450026 (paperback) | ISBN
    9781499450040 (ebook)
Subjects: LCSH: Artificial intelligence–Juvenile literature. | Natural
    language processing (Computer science)–Juvenile literature. |
    ChatGPT–Juvenile literature.
Classification: LCC Q335.4 .H366 2025  (print) | LCC Q335.4  (ebook) | DDC
    006.3–dc23/eng/20240628
LC record available at https://lccn.loc.gov/2024026548
LC ebook record available at https://lccn.loc.gov/2024026549

Manufactured in the United States of America



Find us on 

# Contents

# Is It Alive?

Artificial intelligence, AI for short, refers to the ability of some computers to solve problems the way people do. Even though AI helps it do things that make it seem as smart as a person, the computer is not really alive or thinking. Instead, a set of algorithms, which are recipes for step-by-step completion of tasks, has been combined into a **complex** program.

Like a human, AI programs can answer questions, write stories, and even create art. It's hard to believe that something that can draw a picture of a fluffy kitten and then describe petting its soft fur is just a series of algorithms. That's why we call this **technology** "artificial intelligence." It seems to be smart and alive, but it's just a computer program.

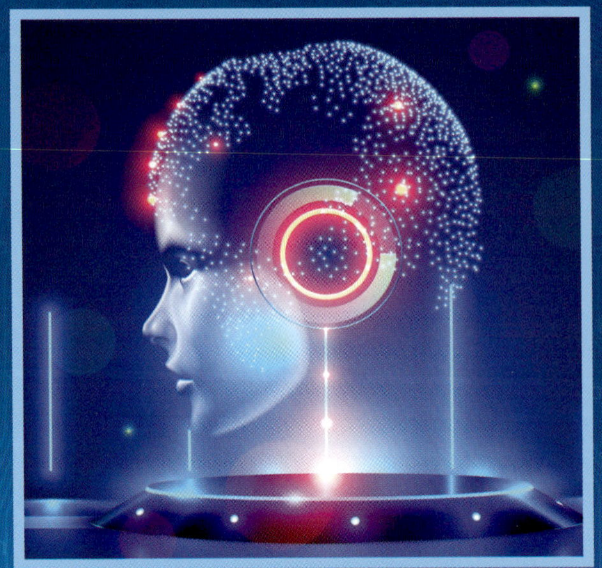

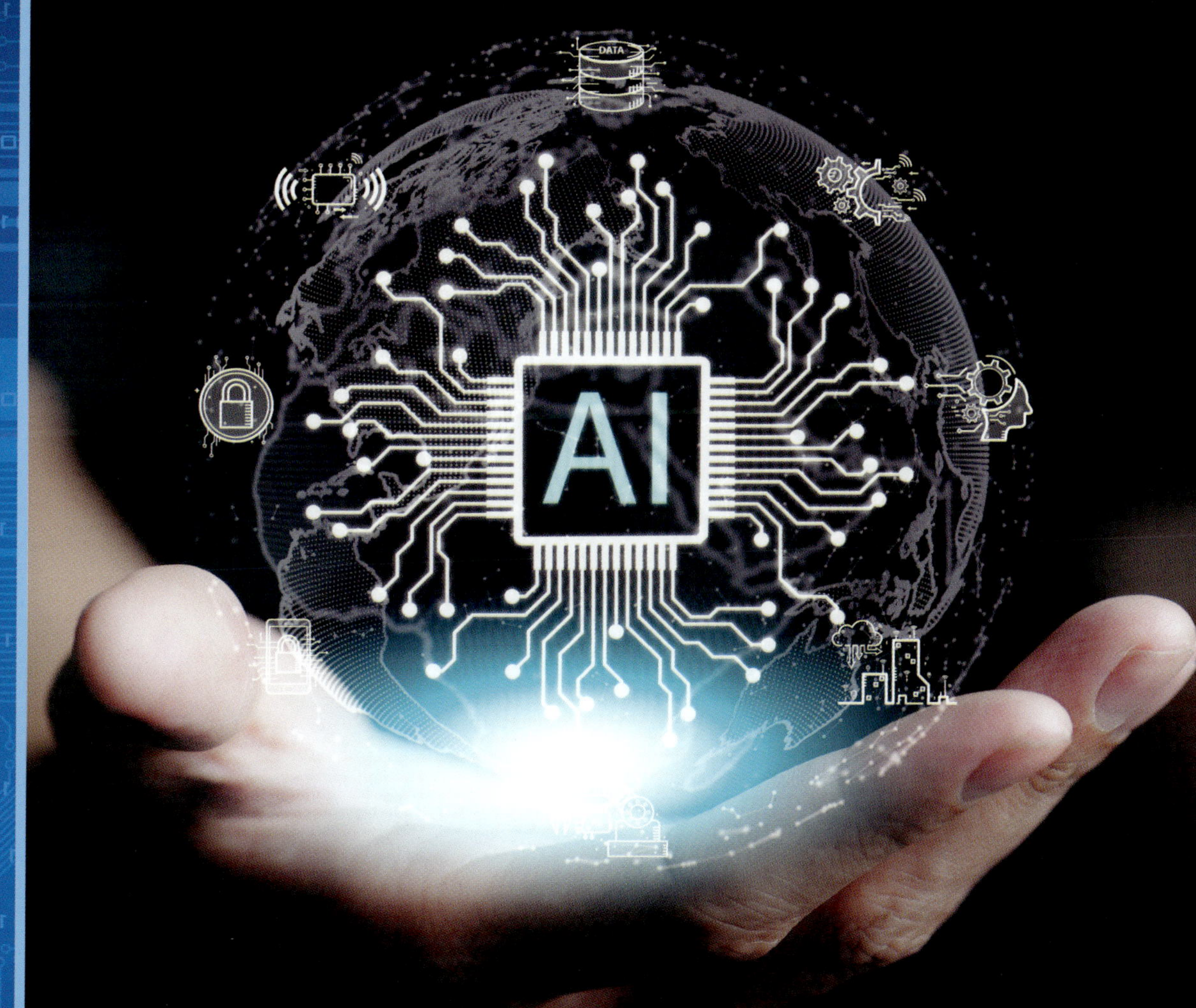

In movies, AI is sometimes shown as something scary that has come to life and wants to harm people. Actually, AI is just an ability some computers or programs have.

# The History of AI

AI is older than you might think. In the 1960s, programmers at the Massachusetts Institute of Technology (MIT) created ELIZA, an early AI program that would parse what humans typed and then respond. Parsing is how computers handle the task of figuring out important words within a longer sentence. This is a part of natural language processing that makes it seem like computers can read and understand languages like a person does.

While the simple parsing and responses of ELIZA wouldn't fool most people into thinking they were talking with a real person, it showed the possibility for improvement. In 1950, Alan Turing had proposed the famous Turing test for determining how humanlike a computer program could become. Though the original ELIZA couldn't pass the Turing test, modern AI has become much better at natural language processing and much better at imitating human responses.

```
Welcome to

              EEEEEE  LL       IIII   ZZZZZZ    AAAAA
              EE      LL        II        ZZ   AA    AA
              EEEEE   LL        II       ZZZ    AAAAAAA
              EE      LL        II      ZZ     AA    AA
              EEEEEE  LLLLLL   IIII  ZZZZZZ     AA    AA

   Eliza is a mock Rogerian psychotherapist.
   The original program was described by Joseph Weizenbaum in 1966.
   This implementation by Norbert Landsteiner 2005.


ELIZA: Is something troubling you ?
YOU:    Men are all alike.
ELIZA: What is the connection, do you suppose ?
YOU:    They're always bugging us about something or other.
ELIZA: Can you think of a specific example ?
YOU:    Well, my boyfriend made me come here.
ELIZA: Is it important to you that your boyfriend made you come here ?
YOU:    He says I'm depressed much of the time.
ELIZA: I am sorry to hear that you are depressed.
YOU:    It's true. I am unhappy.
ELIZA: Can you explain what made you unhappy ?
YOU:
```

# Breaking the Code

There are different **versions** of the Turing test, but most of them involve a person asking questions of two unknown individuals. The test asks if the person can correctly identify which of the unknown individuals is a human and which is a computer. Some researchers claim that ChatGPT can pass the Turing test.

# How Computers Learn

Though they aren't alive, computer programs can be taught through a process called **machine learning**. By giving an AI enough right and wrong examples, the AI can get better at its work. When websites ask, "Are you a human?" and then make you click on boxes that have a bus or a bicycle, you may be also helping a machine learn by providing more examples.

What makes modern AI programs so much more powerful than early programs like ELIZA is that they have been trained on enormous sets of data. Imagine an AI reading every book ever written! The result is a powerful new program called a large language model, or LLM. Because the LLM has so many examples of human writing, it has learned to write more like a person.

In 2000, Pandora launched as an AI-powered streaming service that had been trained on all kinds of different music **genres** to suggest songs for users.


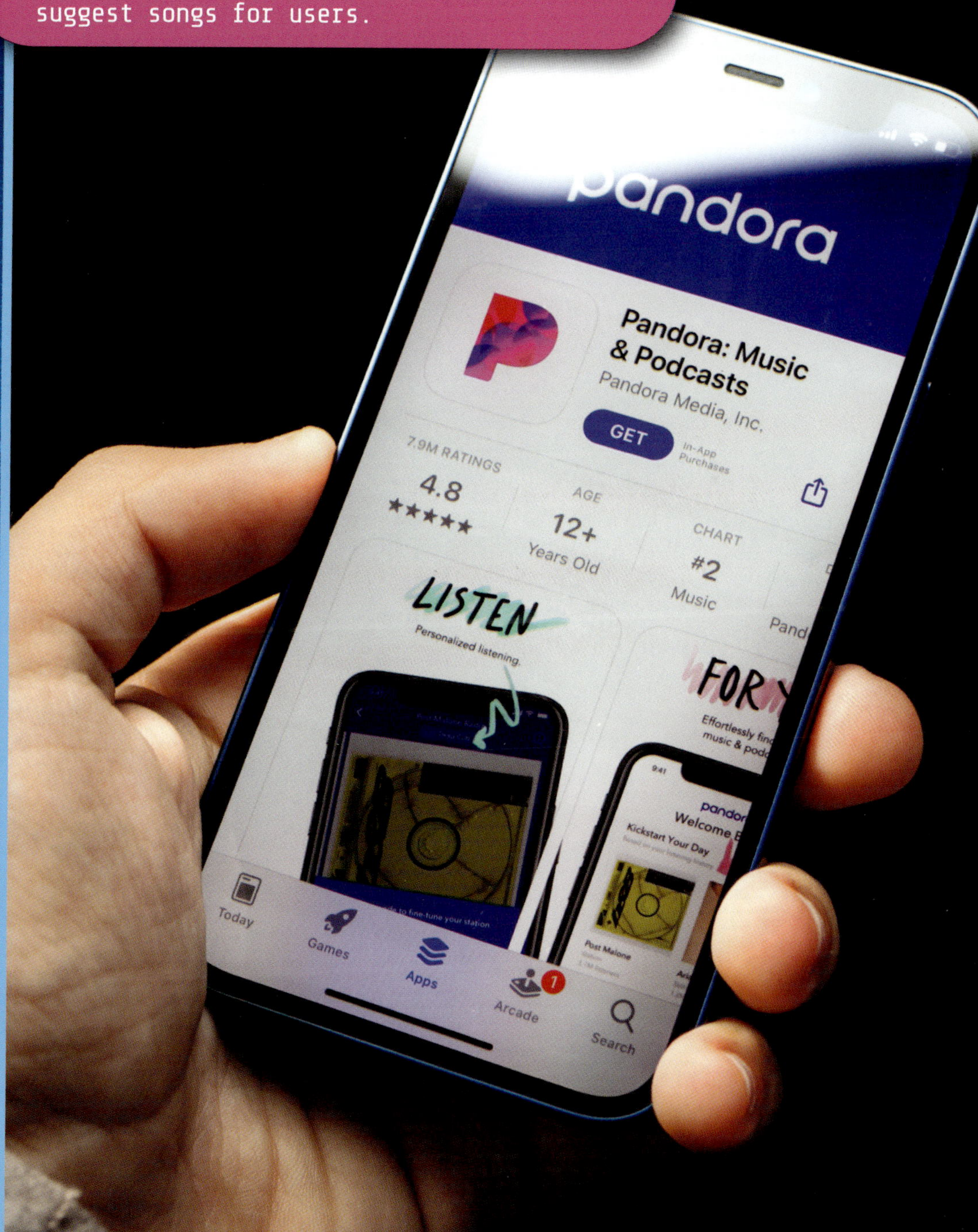

# The Reality of Generative AI

The ability of modern AI to write, or generate, new text that seems as if it was written by a person is why it's called generative AI. This isn't really the best word for what's happening. When we talk about something being generated, we suggest that it was created from nothing and that it's a new thing. Generative AI isn't really creating new things. Instead, it is using the enormous set of examples from a LLM to **mimic** human writing.

Remember, AI doesn't really think. What it does is use lots and lots of complicated math to **predict** what words would likely come next in a sentence or what lines would come next in a drawing. Those predictions are based on all the examples stored in the LLM.

## Breaking the Code

Generative AI uses math to mimic human writing by checking what word might come next in a sentence. This is done by checking the percentage chance of many different word combinations. One interesting thing is that the AI doesn't always pick the next word with the highest percentage.

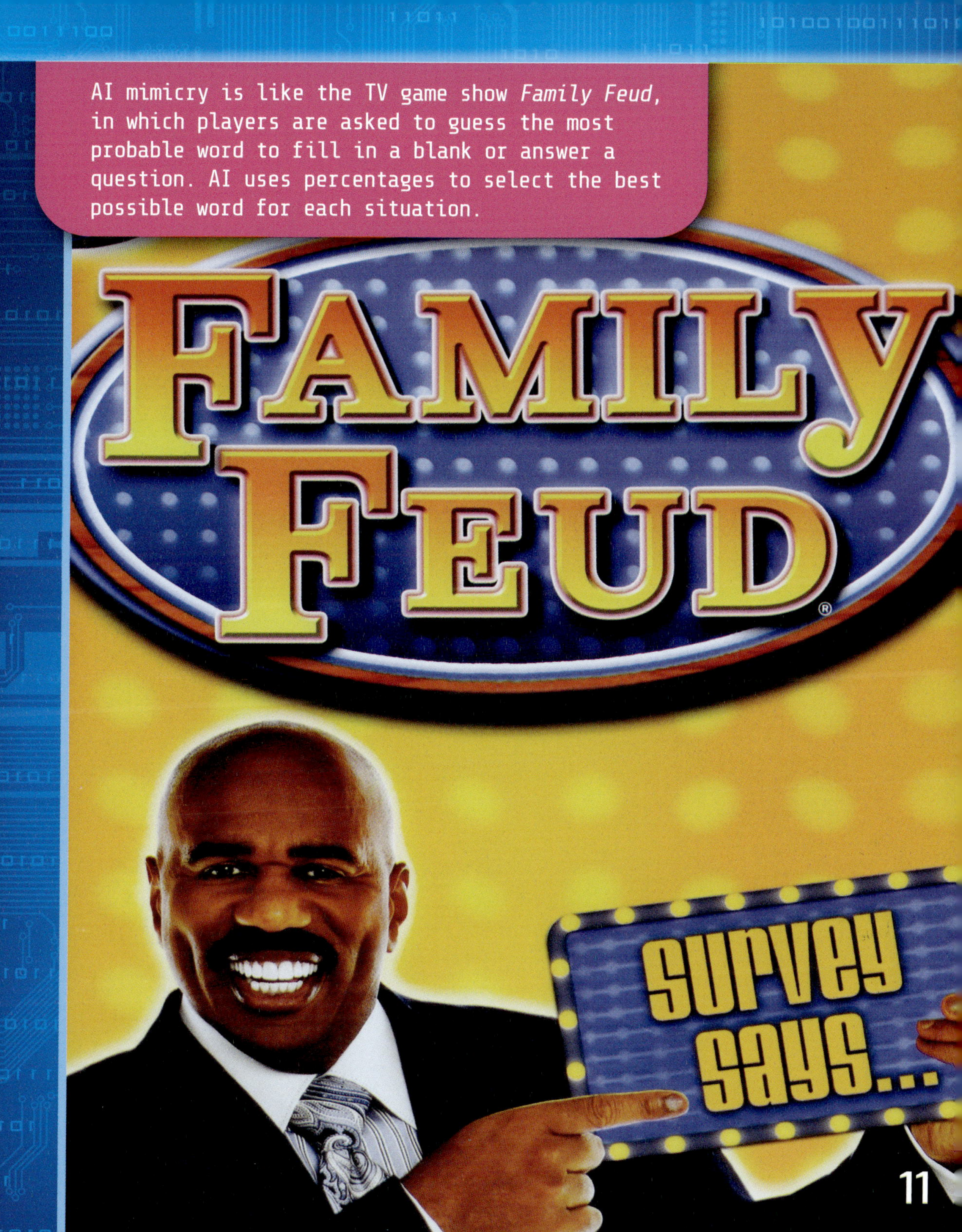


AI mimicry is like the TV game show Family Feud, in which players are asked to guess the most probable word to fill in a blank or answer a question. AI uses percentages to select the best possible word for each situation.

FAMILY FEUD

survey says...

# Ethical Issues with AI

The problem with AI imitating what it finds in an LLM is that sometimes the AI learns things that aren't true because of the examples it has seen. If an AI was only shown pictures of black-and-white-spotted Holstein cows, then it may not realize there are other types of cows as well. This means that the AI can be taught **bias**.

This is an **ethical** issue with the use of AI. The risk of bias in AI training means that humans should always be involved in reviewing output from an AI. This is called HITL, or human-in-the-loop. While the best way to address bias is to have the most **diverse** set of training data possible, HITL can also help detect and fix any bias issues.

Modern airplanes can mostly fly themselves, but there are still human pilots in the cockpit to keep watch on the computers and act in case problems arise.

# Robot Vacuum or Electric Bicycle?

People at the U.S. Department of Education came up with a great way to describe HITL, the risk of bias, and other problems that can arise from using AI. In 2023, as generative AI was just starting to be used, they wrote that students and teachers should think about using AI like an electric bicycle and not like a robot vacuum cleaner.

What they meant is that a student shouldn't give up all their responsibility for doing work and expect that a robot vacuum cleaner will do their chores for them. It may miss spots on the floor or get stuck in a corner and not complete the task! Instead, you should use AI like an electric bicycle. You are still in control but can use the motor on the bicycle to help you get up a hill.

Don't give up your power and control over AI. Make sure you are still the one driving the bicycle even if you use AI for help getting up a hill.

# The New Age of Generative AI

Computer programmers have been working on improving AI for many years. In 1997, an AI called Deep Blue developed by IBM was the first program to win a game of chess against a person—world chess champion Garry Kasparov. Then, in 2011, IBM's Watson AI won the game show *Jeopardy!* against two human champions.

The next big leap forward was the release of ChatGPT 3 by OpenAI on November 30, 2022. ChatGPT isn't a supercomputer limited to only a few users. It was made freely available for anyone to explore and use. People discovered a complex AI that was able to write and interact at a level of apparent humanity that hadn't been seen before. Since 2022, companies have released new versions of ChatGPT and many other generative AI programs.

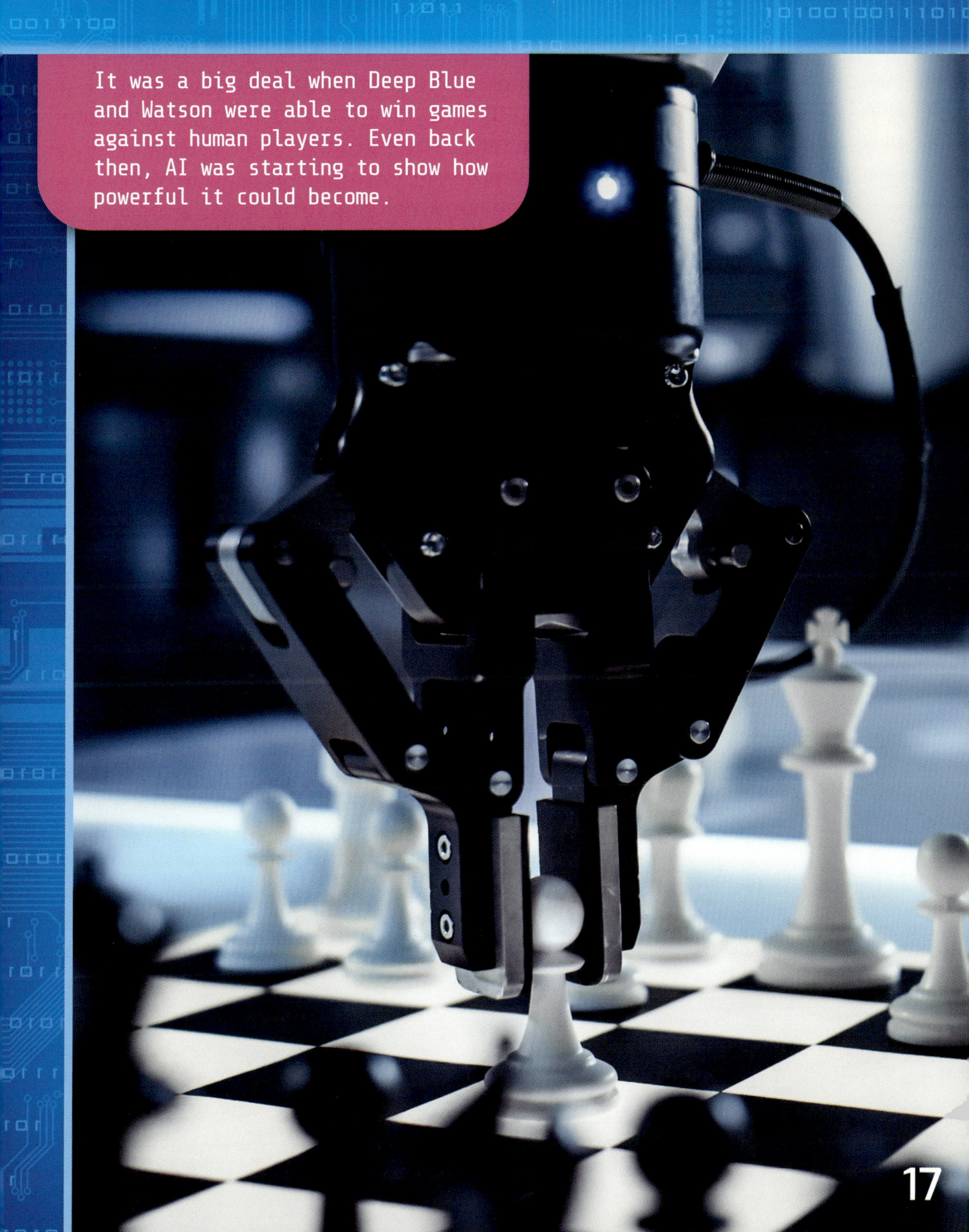
It was a big deal when Deep Blue and Watson were able to win games against human players. Even back then, AI was starting to show how powerful it could become.

# Generative AI Goes to Work

Generative AI has already had a large impact on how we learn and work. Teachers are changing their assignments to avoid students using AI to cheat instead of doing the work themselves. However, using AI to do work is going to be a big part of business moving forward. Some companies are looking to the technology to help their workers be more productive.

To get the best work out of AI, though, you will need to know how to write an effective prompt. Prompts are the directions, similar to code, that you give the AI to tell it what you want it to do. The best prompts describe the problem you are trying to solve and give clear directions. Being able to write good prompts will be an important job skill for future workers.

## Breaking the Code

Researchers from MIT found that people who used ChatGPT to help them complete writing tasks for their jobs ended up finishing their work about 40 percent faster than those who didn't use AI. More importantly, those who got help from AI scored about 18 percent better at writing. This shows that AI can be helpful in the workplace.

Some schools use AI to try to figure out if student work was written by a human or an AI. However, these AI-checking programs are often wrong.

# Using AI to Get Creative

AI isn't just about work. Using the same process of training with a LLM, new AI programs like DALL-E, Midjourney, and Stable Diffusion have become better at creating artwork. Again, this isn't new. Harold Cohen began working on the AARON art AI in the 1970s. By 1995, AARON was able to create art in full color.

Modern image AI programs can be used to create things like artwork for advertisements and even what appear to be photographs of impossible things. Ethical users of image AI, however, should never use it to create **deepfake** images that look like real people in bad situations. As with AI writing, it is important to carefully review AI images. They can include **oddities**, such as a hand with too many fingers, known as hallucinations.

At first glance, this might look like a real photograph of a father and son. However, it includes an AI hallucination!

# Where Is AI Going?

One big possibility for the future of AI is artificial general intelligence, or AGI, that could complete almost any task as well or better than a human. As amazing as generative AI seems at first, when you use it for a while, it's easy to see that it isn't really thinking like a person. AI can make up fake information, show hallucinations, and still get even simple math questions wrong.

Some computer scientists think we may be able to achieve AGI within 20 or 30 years, but others think it will take at least 100 years of work. What most AI experts agree on, though, is that we need to be careful to address concerns like bias and deepfakes so that AI can be a positive part of our computing future.

# Glossary

**bias:** A way of thinking that favors one way of feeling or acting over any other. In AI, results produced by algorithms that are unfair or distorted from reality.

**complex:** Not easy to understand or explain; having many parts.

**deepfake:** An image or recording that has been altered to misrepresent someone as doing or saying something that was not actually done or said.

**diverse:** Having many different types, forms, or ideas.

**ethical:** Following accepted rules of conduct, being the right thing to do.

**genre:** A type or category of literary, musical, or artistic creation, such as western novels or landscape paintings.

**machine learning:** A branch of computer science that focuses on using data and algorithms to enable AI to imitate the way that humans learn, gradually improving its accuracy.

**mimic:** To copy the way something looks, acts, or sounds.

**oddity:** Something that is out of the ordinary.

**predict:** To guess what will happen in the future based on facts or knowledge.

**technology:** A method that uses science to solve problems and the tools used to solve those problems.

**version:** A form of something that is different from the ones that came before it.

# Index